Alembic Club Reprints—No. 5

EXTRACTS FROM MICROGRAPHIA: OR SOME PHYSIOLOGICAL DESCRIPTIONS OF MINUTE BODIES MADE BY MAGNIFYING GLASSES, WITH OBSERVATIONS AND INQUIRIES THEREUPON.

BY

R. HOOKE. FELLOW OF THE ROYAL SOCIETY

(1665)

EDINBURGH
THE ALEMBIC CLUB

CHICAGO
THE UNIVERSITY OF CHICAGO PRESS
1906

PREFACE.

HOOKE'S "Micrographia" is a remarkable book in several respects, but in none more strikingly so than on account of the explanation which it contains of its author's clear views respecting the theory of combustion. This explanation occurs quite unexpectedly in Observation XVI., entitled "Of Charcoal, or Burnt Vegetables," and it forms by far the most important portion of the book from the chemical standpoint; although matters of minor chemical interest are mentioned in several of the other Observations. It is a matter for the greatest regret that Hooke did not detail in Observation XVI. the experiments from which his theoretical ideas concerning combustion were deduced; and also that he never carried out his expressed intention of dealing fully with the whole subject in a separate treatise.

A just appreciation of the author's individuality, as well as of his intentions in publishing his Micrographia, can scarcely be obtained without perusing his preface, which contains a number of shrewd and enlightened ideas.

The following pages contain the greater part of the preface to Micrographia, Observations VIII. and XVI. in full, and a few shorter extracts. The old-fashioned spelling of the original has been retained.

L. D.

MICROGRAPHIA.

EXTRACTS FROM THE PREFACE.

IT is the great prerogative of Mankind above other Creatures, that we are not only able to *behold* the works of Nature, or barely to *sustein* our lives by them, but we have also the power of *considering*, *comparing*, *altering*, *assisting*, and *improving* them to various uses. And as this is the peculiar priviledge of humane Nature in general, so is it capable of being so far advanced by the helps of Art, and Experience, as to make some Men excel others in their Observations, and Deductions, almost as much as they do Beasts. By the addition of such *artificial Instruments* and *methods*, there may be, in some manner, a reparation made for the mischiefs, and imperfection, mankind has drawn upon it self, by negligence, and intemperance, and a wilful and superstitious deserting the Prescripts and Rules of Nature, whereby every man, both from a deriv'd corruption, innate and born with him, and from his breeding and converse with men, is very subject to slip into all sorts of errors.

The only way which now remains for us to recover some degree of those former perfections, seems to be, by rectifying the operations of the *Sense*, the *Memory*, and *Reason*, since upon the evidence, the *strength*, the *integrity*, and the *right correspondence* of all these, all the light, by which our actions are to be guided, is to be renewed, and all our command over things is to be establisht.

It is therefore most worthy of our consideration, to recollect their several defects, that so we may the better understand how to supply them, and by what assistances we may *inlarge* their power, and *secure* them in performing their particular duties.

As for the actions of our *Senses*, we cannot but observe them to be in many particulars much outdone by those of other Creatures, and when at best, to be far short of the perfection they seem capable of: And these infirmities of the Senses arise from a double cause, either from the *disproportion of the Object to the Organ*, whereby an infinite number of things can never enter into them, or else from *error in the Perception*, that many things, which come within their reach, are not received in a right manner.

The like frailties are to be found in the *Memory;* we often let many things *slip away* from us, which deserve to be retain'd; and of those which we treasure up, a great part is either *frivolous* or *false;* and if good, and substantial, either in tract of time *obliterated*, or at best so *overwhelmed* and buried under more frothy notions, that when there is need of them, they are in vain sought for.

The two main foundations being so deceivable, it is no wonder, that all the succeeding works which we build upon them, of arguing, concluding, defining, judging, and all the other degrees of Reason, are lyable to the same imperfection, being, at best, either vain, or uncertain: So that the errors of the *understanding* are answerable to the two other, being defective both in the quantity and goodness of its knowledge; for the limits, to which our thoughts are confined, are small in respect of the vast extent of Nature it self; some parts of it are *too large* to be comprehended, and some *too little* to be perceived. And from thence it must follow, that not having a full sensation of the Object, we must be very lame and imperfect in our conceptions about it, and in all the propositions which we build upon it; hence we often take the ***shadow*** of things for the ***substance***, small ***appearances*** for good ***similitudes***, ***similitudes*** for ***definitions;*** and even many of those, which we think to be the most solid definitions, are rather expressions of our own misguided apprehensions then of the true nature of the things themselves.

The effects of these imperfections are manifested in different ways, according to the temper and disposition of the several minds of men, some they incline to ***gross ignorance*** and stupidity, and others to a ***presumptuous imposing*** on other mens Opinions, and a ***confident dogmatizing*** on matters, whereof there is no assurance to be given.

Thus all the uncertainty, and mistakes of humane actions, proceed either from the narrowness and wandring of our ***Senses***, from the slipperiness or delusion of our ***Memory***, from the confinement or rashness of our ***Understanding***, so that 'tis no wonder, that our power over natural causes and effects is so slowly improv'd, seeing we are not only to contend with the obscurity and ***difficulty of the things*** whereon we work and think, but even the ***forces of our own minds*** conspire to betray us.

These being the dangers in the process of humane Reason, the remedies of them all can only proceed from the ***real***, the ***mechanical***, the ***experimental*** Philosophy, which has this advantage over the Philosophy of ***discourse*** and ***disputation***, that whereas that chiefly aims at the subtilty of its Deductions and Conclusions, without much regard to the first ground-work, which ought to be well laid on the Sense and Memory; so this intends the right ordering of them all, and the making them serviceable to each other.

The first thing to be undertaken in this weighty work, is a *watchfulness over the failings* and an *inlargement of the dominion*, of the Senses.

To which end it is requisite, first, That there should be a *scrupulous* choice, and a *strict examination*, of the reality, constancy, and certainty of the Particulars that we admit: This is the first rise whereon truth is to begin, and here the most severe, and most impartial diligence, must be imployed; the storing up of all, without any regard to evidence or use, will only tend to darkness and confusion. We must not therefore esteem the riches of our Philosophical treasure by the *number* only, but chiefly by the *weight*; the most *vulgar* Instances are not to be neglected, but above all, the most *instructive* are to be entertain'd; the footsteps of Nature are to be trac'd, not only in her *ordinary course*, but when she seems to be put to her shifts, to make many *doublings* and *turnings*, and to use some kind of art in indeavouring to avoid our discovery.

The next care to be taken, in respect of the Senses, is a supplying of their infirmities with *Instruments*, and, as it were, the adding of *artificial Organs* to the *natural*; this in one of them has been of late years accomplisht with prodigious benefit to all sorts of useful knowledge, by the invention of Optical Glasses. By the means of *Telescopes*, there is nothing so *far distant* but may be represented to our view; and by the help of *Microscopes*, there is nothing so *small*, as to escape our inquiry; hence there is a new visible World discovered to the understanding. By this means the Heavens are open'd, and a vast number of new Stars, and new Motions, and new Productions appear in them, to which all the antient Astronomers were utterly Strangers. By this the Earth it self, which lyes so neer us, under our feet, shews quite a new thing to us, and in every *little particle* of its matter, we now behold almost as great a variety of Creatures, as we were able before to reckon up in the whole *Universe* it self.

It seems not improbable, but that by these helps the subtilty of the composition of Bodies, the structure of their parts, the various texture of their matter, the instruments and manner of their inward motions, and all the other possible appearances of things, may come to be more fully discovered; all which the antient *Peripateticks* were content to comprehend in two general and (unless further explain'd) useless words of *Matter* and *Form*. From whence there may arise many admirable advantages, towards the increase of the *Operative*, and the *Mechanick* Knowledge, to which this Age seems so much inclined, because we may perhaps be inabled to discern all the secret workings of Nature, almost in the same manner as we do those that are the productions of Art, and are manag'd by Wheels, and Engines, and Springs, that were devised by humane Wit.

In this kind I here present to the World my imperfect Indeavours; which though they shall prove no other way considerable, yet, I hope, they may be in some measure useful to the main Design of a *reformation* in Philosophy, if it be only by shewing, that there is not so much requir'd towards it, any strength of *Imagination*, or exactness of *Method*, or depth of *Contemplation* (though the addition of these, where they can be had, must needs produce a much more perfect composure) as a *sincere Hand*, and a *faithful Eye*, to examine, and to record, the things themselves as they appear.

And I beg my Reader, to let me take the boldness to assure him, that in this present condition of knowledge, a man so qualified, as I have indeavoured to be, only with resolution, and integrity, and plain intentions of imploying his *Senses* aright, may venture to compare the reality and the usefulness of his services, towards the true Philosophy, with those of other men, that are of much stronger, and more acute *speculations*, that shall not make use of the same method by the Senses.

The truth is, the Science of Nature has been already too long made only a work of the *Brain* and the *Fancy:* It is now high time that it should return to the plainness and soundness of *Observations* on *material* and *obvious* things. It is said of great Empires, That *the best way to preserve them from decay, is to bring them back to the first Principles, and Arts, on which they did begin.* The same is undoubtedly true in Philosophy, that by wandring far away into *invisible Notions*, has almost quite destroy'd it self, and it can never be recovered, or continued, but by returning into the same *sensible paths*, in which it did at first proceed.

If therefore the Reader expects from me any infallible Deductions, or certainty of *Axioms*, I am to say for my self, that those stronger Works of Wit and Imagination are above my weak Abilities; or if they had not been so, I would not have made use of them in this present Subject before me: Whereever he finds that I have ventur'd at any small Conjectures, at the causes of the things that I have observed, I beseech him to look upon them only as *doubtful Problems*, and *uncertain ghesses*, and not as unquestionable Conclusions, or matters of unconfutable Science; I have produced nothing here, with intent to bind his understanding to an *implicit* consent; I am so far from that, that I desire him, not absolutely to rely upon these Observations of my eyes, if he finds them contradicted by the future Ocular Experiments of sober and impartial Discoverers.

As for my part, I have obtained my end, if these my small Labours shall be thought fit to take up some place in the large stock of *natural Observations*, which so many hands are busie in providing. If I have contributed the *meanest foundations* whereon others may raise nobler *Superstructures*, I am abundantly satisfied; and all my ambition is, that I may serve to the great Philosophers of this Age, as the makers and the grinders of my Glasses did to me; that I may prepare and furnish them with some *Materials*, which they may afterwards *order* and *manage* with better skill, and to far greater advantage.

The next remedies in this universal cure of the Mind are to be applyed to the *Memory*, and they are to consist of such Directions as may inform us, what things are best to be *stor'd up* for our purpose, and which is the best way of so *disposing* them, that they may not only be *kept in safety*, but ready and convenient, to be at any time *produc'd* for use, as occasion shall require. But I will not here prevent my self in what I may say in another Discourse, wherein I shall make an attempt to propose some Considerations of the manner of compiling a Natural and Artificial History, and of so ranging and registring its Particulars into Philosophical Tables, as may make them most useful for the raising of *Axioms* and *Theories*.

The last indeed is the most *hazardous* Enterprize, and yet the most *necessary*; and that is, to take such care that the *Judgment* and the *Reason* of Man (which is the third Faculty to be repair'd and improv'd) should receive such assistance, as to avoid the dangers to which it is by nature most subject. The Imperfections, which I have already mention'd, to which it is lyable, do either belong to the *extent*, or the *goodness* of its knowledge; and here the difficulty is the greater, least that which may be thought a *remedy* for the one should prove *destructive* to the other, least by seeking to inlarge our Knowledge, we should render it weak and uncertain; and least by being too scrupulous and exact about every Circumstance of it, we should confine and streighten it too much.

In both these the middle wayes are to be taken, nothing is to be *omitted*, and yet every thing to pass a *mature deliberation* : No *Intelligence* from Men of all Professions, and quarters of the World, to be *slighted*, and yet all to be so *severely examin'd*, that there remain no room for doubt or instability; much *rigour* in admitting, much *strictness* in comparing, and above all, much *slowness* in debating, and *shyness* in determining, is to be practised. The *Understanding* is to *order* all the inferiour services of the lower Faculties ; but yet it is to do this only as a *lawful Master*, and not as a *Tyrant*. It must not *incroach* upon their Offices, nor take upon it self the employments which belong to either of them. It must *watch* the irregularities of the Senses, but it must not go before them, or *prevent* their information. It must *examine*, *range*, and *dispose* of the bank which is laid up in the Memory ; but it must be sure to make *distinction* between the *sober* and *well collected heap*, and the *extravagant Idea's*, and *mistaken Images*, which there it may sometimes light upon. So many are the *links*, upon which the true Philosophy depends, of which, if any one be *loose*, or *weak*, the whole *chain* is in danger of being dissolv'd ; it is to *begin* with the Hands and Eyes, and to *proceed* on through the Memory, to be *continued* by the Reason ; nor is it to stop there, but to *come about* to the Hands and Eyes again, and so, by a *continual passage round* from one Faculty to another, it is to be maintained in life and strength, as much as the body of man is by the *circulation* of the blood through the several parts of the body, the Arms, the Feet, the Lungs, the Heart, and the Head.

If once this method were followed with diligence and attention, there is nothing that lyes within the power of human Wit (or which is far more effectual) of human Industry, which we might not compass; we might not only hope for Inventions to equalize those of *Copernicus*, *Galileo*, *Gilbert*, *Harvy*, and of others, whose Names are almost lost, that were the Inventors of *Gun-powder*, the *Seamans Compass*, *Printing*, *Etching*, *Graving*, *Microscopes*, *&c.* but multitudes that may far exceed them: for even those discoveries seem to have been the products of some such method, though but imperfect; What may not be therefore expected from it if thoroughly prosecuted? *Talking* and *contention of Arguments* would soon be turn'd into *labours;* all the fine *dreams* of Opinions, and *universal metaphysical natures*, which the luxury of subtil Brains has devis'd, would quickly vanish, and give place to *solid Histories*, *Experiments* and *Works*. And as at first, mankind *fell* by *tasting* of the forbidden Tree of Knowledge, so we, their Posterity, may be in part *restor'd* by the same way, not only by *beholding* and *contemplating*, but by *tasting* too those fruits of Natural knowledge, that were never yet forbidden.

From hence the World may be assisted with *variety* of Inventions, *new* matter for Sciences may be *collected*, the *old improv'd*, and their *rust* rubb'd away; and as it is by the benefit of Senses that we receive all our Skill in the works of Nature, so they also may be wonderfully benefited by it, and may be guided to an easier and more exact performance of their Offices; 'tis not unlikely, but that we may find out wherein our Senses are deficient, and as easily find wayes of repairing them.

The Indeavours of Skilful men have been most conversant about the assistance of the Eye, and many noble Productions have followed upon it; and from hence we may conclude, that there is a way open'd for advancing the operations, not only of all the other Senses, but even of the Eye it self; that which has been already done ought not to content us, but rather to incourage us to proceed further, and to attempt greater things in the same and different wayes.

'Tis not unlikely, but that there may be yet invented several other helps for the eye, as much exceeding those already found, as those do the bare eye, such as by which we may perhaps be able to discover *living Creatures* in the Moon, or other Planets, the *figures* of the compounding Particles of matter, and the particular *Schematisms* and *Textures* of Bodies.

And as *Glasses* have highly promoted our *seeing*, so 'tis not improbable, but that there may be found many *Mechanical Inventions* to improve our other Senses, of *hearing*, *smelling*, *tasting*, *touching*. 'Tis not impossible to hear a *whisper* a *furlongs* distance, it having been already done; and perhaps the nature of the thing would not make it more impossible, though that furlong should be ten times multiply'd. And though some famous Authors have affirm'd it impossible to hear through the *thinnest plate* of *Muscovy-glass*; yet I know a way, by which 'tis easie enough to hear one speak through a *wall a yard thick*. It has not been yet thoroughly examin'd, how far *Otocousticons* may be improv'd, nor what other wayes there may be of *quickning* our hearing, or *conveying* sound through *other bodies* then the *Air*: for that that is not the only *medium*, I can assure the Reader, that I have, by the help of a *distended wire*, propagated the sound to a very considerable distance in an *instant*, or with as seemingly quick a motion as that of light, at least, incomparably swifter then that, which at the same time was propagated through the Air; and this not only in a straight line, or direct, but in one bended in many angles.

Nor are the other three so perfect, but that *diligence*, *attention*, and many *mechanical contrivances*, may also highly improve them. For since the sense of *smelling* seems to be made by the *swift passage* of the *Air* (*impregnated* with the steams and *effluvia* of several odorous Bodies) through the grisly *meanders* of the Nose whose surfaces are *cover'd* with a very sensible *nerve*, and *moistned* by a *transudation* from the *processus mamillares* of the Brain, and some adjoyning *glandules*, and by the moist *steam* of the *Lungs*, with a Liquor convenient for the reception of those *effluvia* and by the adhesion and mixing of those steams with that liquor, and thereby affecting the nerve, or perhaps by insinuating themselves into the juices of the brain, after the same manner, as I have in the following Observations intimated, the parts of Salt to pass through the skins of Effs, and Frogs. Since, I say, smelling seems to be made by some such way, 'tis not improbable, but that some contrivance, for making a great quantity of Air pass quick through the Nose, might as much promote the sense of smelling, as the any wayes hindring that passage does dull and destroy it. Several tryals I have made, both of hindring and promoting this sense, and have succeeded in some according to expectation; and indeed to me it seems capable of being improv'd, for the judging of the constitutions of many Bodies. Perhaps we may thereby also judge (as other Creatures seem to do) what is wholsome, what poyson; and in a word, what are the specifick properties of Bodies.

There may be also some other mechanical wayes found out, of sensibly perceiving the *effluvia* of Bodies; several Instances of which, were it here proper, I could give of Mineral steams and exhalations; and it seems not impossible, but that by some such wayes improved, may be discovered, what Minerals lye buried under the Earth, without the trouble to dig for them; some things to confirm this Conjecture may be found in *Agricola*, and other Writers of Minerals, speaking of the Vegetables that are apt to thrive, or pine, in those steams.

Whether also those steams, which seem to issue out of the Earth, and mix with the Air (and so to precipitate some *aqueous* Exhalations, wherewith 'tis impregnated) may not be by some way detected before they produce the effect, seems hard to determine; yet something of this kind I am able to discover, by an Instrument I contriv'd to shew all the minute variations in the pressure of the Air; by which I constantly find, that before, and during the time of rainy weather, the pressure of the Air is less, and in *dry weather*, but especially when an *Eastern Wind* (which having past over vast tracts of Land is heavy with Earthy Particles) blows, it is much more, though these changes are varied according to very odd Laws.

* * *

But this is but one way of discovering the *effluvia* of the Earth mixt with the Air; there may be perhaps many others, witness the *Hygroscope*, an Instrument whereby the watery steams volatile in the Air are discerned, which the Nose it self is not able to find. This I have describ'd in the following Tract in the Description of the Beard of a wild Oat. Others there are, may be discovered both by the Nose, and by other wayes also. Thus the *smoak* of burning *Wood* is *smelt*, *seen*, and sufficiently *felt* by the eyes: The *fumes* of burning *Brimstone* are *smelt* and discovered also by the destroying the Colours of Bodies, as by the *whitening of a red Rose*: And who knows, but that the Industry of man, following this method, may find out wayes of improving this sense to as great a degree of perfection as it is in any Animal, and perhaps yet higher.

'Tis not improbable also, but that our *taste* may be very much improv'd, either by *preparing* our tast for the

Body, as, after eating *bitter* things, *Wine*, or other *Vinous liquors*, are more sensibly tasted; or else by *preparing* Bodies for our tast; as the dissolving of Metals with acid Liquors, make them tastable, which were before altogether insipid; thus *Lead* becomes *sweeter* then Sugar, and *Silver* more *bitter* then Gall, *Copper* and *Iron* of most *loathsome* tasts. And indeed the business of this sense being to discover the presence of dissolved Bodies in Liquors put on the Tongue, or in general to discover that a fluid body has some solid body dissolv'd in it, and what they are; whatever contrivance makes this discovery improves this sense. In this kind the mixtures of Chymical Liquors afford many Instances; as the sweet Vinegar that is impregnated with Lead may be discovered to be so by the affusion of a little of an *Alcalizate solution*: The bitter liquor of *Aqua fortis* and *Silver* may be discover'd to be charg'd with that Metal, by laying in it some plates of Copper: 'Tis not improbable also, but there may be multitudes of other wayes of discovering the parts dissolv'd, or dissoluble in liquors; and what is this discovery but a kind of *secundary tasting.*

'Tis not improbable also, but that the sense of *feeling* may be highly improv'd, for that being a sense that judges of the more *gross* and *robust motions* of the *Particles* of *Bodies*, seems capable of being improv'd and assisted very many wayes. Thus for the distinguishing of *Heat* and *Cold*, the *Weather-glass* and *Thermometer*, which I have describ'd in this following Treatise, do exceedingly perfect it; by each of which the least variations of heat or cold, which the most Acute sense is not able to distinguish, are manifested. This is oftentimes further promoted also by the help of *Burning-glasses*, and the like, which collect and unite the radiating heat. Thus the *roughness* and *smoothness* of a Body is made much more sensible by the help of a *Microscope*, then by the most *tender* and *delicate Hand*. Perhaps, a Physitian might, by several other *tangible* proprieties, discover the constitution of a Body as well as by the *Pulse*. I do but instance in these, to shew what possibility there may be of many others, and what probability and hopes there were of finding them, if this method were followed; for the Offices of the five Senses being to detect either the *subtil* and *curious Motions* propagated through all *pellucid* or perfectly *homogeneous* Bodies; Or the more *gross* and *vibrative Pulse* communicated through the *Air* and all other convenient *mediums*, whether fluid or solid: Or the *effluvia* of Bodies *dissolv'd* in the *Air;* Or the *particles* of bodies *dissolv'd* or *dissoluble* in *Liquors*, or the more *quick* and *violent shaking motion* of *heat* in all or any of these: whatsoever does any wayes promote any of these kinds of *criteria*, does afford a way of improving some one sense. And what a multitude of these would a diligent Man meet with in his inquiries? And this for the helping and promoting the *sensitive faculty* only.

Next, as for the *Memory*, or *retentive faculty*, we may be sufficiently instructed from the *written Histories* of *civil actions*, what great assistance may be afforded the Memory, in the committing to writing things observable in *natural operations*. If a Physitian be therefore accounted the more able in his Faculty, because he has had long experience and practice, the remembrance of which, though perhaps very imperfect, does regulate all his after actions: What ought to be thought of that man, that has not only a perfect *register* of his own experience, but is grown *old* with the experience of many hundreds of years, and many thousands of men.

And though of late, men, beginning to be sensible of this convenience, have here and there registred and printed some few *Centuries*, yet for the most part they are set down very lamely and imperfectly, and, I fear, many times not so truly, they seeming, several of them, to be design'd more for *Ostentation* then *publique use*: For, not to instance, that they do, for the most part, omit those Experiences they have made, wherein their Patients have miscarried, it is very easie to be perceiv'd, that they do all along *hyperbolically extol* their own Prescriptions, and vilifie those of others. Notwithstanding all which, these kinds of Histories are generally esteem'd useful, even to the ablest Physitian.

What may not be expected from the *rational* or *deductive Faculty* that is furnisht with such *Materials*, and those so readily *adapted*, and rang'd for use, that in a moment, as 'twere, thousands of Instances, serving for the *illustration*, *determination*, or *invention*, of almost any inquiry, may be *represented* even to the sight? How neer the nature of *Axioms* must all those *Propositions* be which are examin'd before so many *Witnesses?* And how difficult will it be for any, though never so subtil an error in Philosophy, to scape from being discover'd, after it has indur'd the *touch*, and so many other *tryals?*

What kind of mechanical way, and physical invention also is there requir'd, that might not this way be found out? The *Invention* of a way to find the *Longitude* of places is easily perform'd, and that to as great *perfection* as is desir'd, or to as great an *accurateness* as the *Latitude* of places can be found at Sea; and perhaps yet also to a greater certainty then that has been hitherto found, as I shall very speedily freely manifest to the world. The way of *flying* in the Air seems principally unpracticable, by reason of the *want of strength* in *humane muscles;* if therefore that could be supplied, it were, I think, easie to make twenty contrivances to perform the office of *Wings*: What Attempts also I have made for the supplying that Defect, and my successes therein, which, I think, are wholly new, and not inconsiderable, I shall in another place relate.

'Tis not unlikely also, but that *Chymists*, if they followed this method, might find out their so much sought for *Alkahest*. What an *universal Menstruum*, which dissolves all sorts of *Sulphureous Bodies*, I have discover'd (which has not been before taken notice of as such) I have shewn in the sixteenth Observation.

What a prodigious variety of Inventions in *Anatomy* has this latter Age afforded, even in our own Bodies, in the very *Heart*, by which we live, and the *Brain*, which is the seat of our knowledge of other things? witness all the excellent Works of *Pecquet*, *Bartholinus*, *Billius*, and many others; and at home, of Doctor *Harvy*, Doctor *Ent*, Doctor *Willis*, Doctor *Glisson*. In *Celestial Observations* we have far exceeded all the Antients, even the *Chaldeans* and *Egyptians* themselves, whose *vast Plains*, *high Towers*, and *clear Air*, did not give them so great advantages over us, as we have over them by our *Glasses*. By the help of which, they have been very much outdone by the famous *Galileo*, *Hevelius*, *Zulichem*; and our own Countrymen, Mr. *Rook*, Doctor *Wren*, and the great Ornament of our Church and Nation, the *Lord Bishop of Exeter*. And to say no more in *Aerial Discoveries*, there has been a wonderful progress made by the *Noble Engine* of *the most Illustrious Mr. Boyle*, whom it becomes me to mention with all honour, not only as my particular Patron, but as the *Patron* of *Philosophy* it self; which he every day *increases* by his *Labours*, and *adorns* by his *Example*.

The good success of all these *great Men*, and many others, and the now seemingly great *obviousness* of most of their and divers other Inventions, which from the beginning of the world have been, as 'twere, trod on, and yet not minded till these last *inquisitive* Ages (an Argument that there may be yet behind multitudes of the like) puts me in mind to recommend such Studies, and the prosecution of them by such methods, to the *Gentlemen* of our Nation, whose *leisure* makes them fit to *undertake*, and the *plenty* of their fortunes to *accomplish*, extraordinary things in this way. And I do not only propose this kind of *Experimental Philosophy* as a matter of high *rapture* and *delight* of the mind, but even as a *material* and *sensible Pleasure*. So vast is the *variety of Objects* which will come under their Inspections, so many *different wayes* there are of *handling* them, so great is the *satisfaction* of *finding* out *new things*, that I dare compare the *contentment* which they will injoy, not only to that of *contemplation*, but even to that which most men prefer of *the very Senses themselves*.

And if they will please to take any incouragement from so mean and so imperfect endeavours as mine, upon my own experience, I can assure them, without arrogance, That there has not been any inquiry or Problem in *Mechanicks*, that I have hitherto propounded to my self, but by a certain method (which I may on some other opportunity explain) I have been able presently to examine the possibility of it; and if so, as easily to excogitate divers wayes of performing it: And indeed it is possible to do as much by *this method* in *Mechanicks*, as by *Algebra* can be perform'd in *Geometry*. Nor can I at all doubt, but that the same method is as applicable to *Physical Enquiries*, and as likely to find and reap thence as plentiful a crop of Inventions; and indeed there seems to be no subject so barren, but may with this good husbandry be highly improv'd.

Toward the prosecution of this method in *Physical Inquiries*, I have here and there *gleaned* up an *handful* of Observations, in the collection of most of which I made use of *Microscopes*, and some other *Glasses* and *Instru-*

ments that improve the sense; which way I have herein taken, not that there are not multitudes of useful and pleasant Observables, yet uncollected, obvious enough without the helps of Art, but only to promote the use of Mechanical helps for the Senses, both in the surveying the already visible World, and for the discovery of many others hitherto unknown, and to make us, with the great Conqueror, to be affected that we have not yet overcome one World when there are so many others to be discovered, every considerable improvement of *Telescopes* or *Microscopes* producing new Worlds and *Terra-Incognita's* to our view.

* * *

What the things are I observ'd, the following descriptions will manifest; in brief, they were either *exceeding small Bodies*, or *exceeding small Pores*, or *exceeding small Motions*, some of each of which the Reader will find in the following Notes, and such, as I presume, (many of them at least) will be *new*, and perhaps not less *strange*: Some *specimen* of each of which Heads the Reader will find in the subsequent delineations, and indeed of some more then I was willing there should be; which was occasioned by my first Intentions to print a much greater number then I have since found time to compleat. Of such therefore as I had, I selected only some few of every Head, which for some particulars seem'd most observable, rejecting the rest as superfluous to the present Design.

What each of the delineated Subjects are, the following descriptions annext to each will inform, of which I shall here, only once for all, add, That in divers of them the Gravers have pretty well follow'd my directions and draughts; and that in making of them, I indeavoured (as far as I was able) first to discover the true appearance, and next to make a plain representation of it. This I mention the rather, because of these kind of Objects there is much more difficulty to discover the true shape, then of those visible to the naked eye, the same Object seeming quite differing, in one position to the Light, from what it really is, and may be discover'd in another. And therefore I never began to make any draught before by many examinations in several lights, and in several positions to those lights, I had discover'd the true form. For it is exceeding difficult in some Objects, to distinguish between a *prominency* and a *depression*, between a *shadow* and a *black stain*, or a *reflection* and a *whiteness in the colour*. Besides, the transparency of most Objects renders them yet much more difficult then if they were *opacous*. The Eyes of a Fly in one kind of light appear almost like a Lattice, drill'd through with abundance of small holes; which probably may be the Reason, why the Ingenious *Dr. Power* seems to suppose them such. In the Sunshine they look like a Surface cover'd with golden Nails; in another posture, like a Surface cover'd with Pyramids; in another with Cones; and in other postures of quite other shapes; but that which exhibits the best, is the Light collected on the Object, by those means I have already describ'd.

And this was undertaken in prosecution of the Design which the *ROYAL SOCIETY* has propos'd to it self. For the Members of the Assembly having before their eyes so many *fatal* Instances of the errors and falshoods, in which the greatest part of mankind has so long wandred, because they rely'd upon the strength of humane Reason alone, have begun anew to correct all *Hypotheses* by sense, as Seamen do their *dead Reckonings* by *Cœlestial Observations;* and to this purpose it has been their principal indeavour to *enlarge* and *strengthen* the *Senses* by *Medicine*, and by such *outward Instruments* as are proper for their particular works. By this means they find some reason to suspect, that those effects of Bodies, which have been commonly attributed to *Qualities*, and those confess'd to be *occult*, are perform'd by the small *Machines* of Nature, which are not to be discern'd without these helps, seeming the meer products of *Motion*, *Figure*, and *Magnitude*; and that the *Natural Textures*, which some call the *Plastick faculty*, may be made in *Looms*, which a greater perfection of Opticks may make discernable by these Glasses; so as now they are no more puzzled about them, then the vulgar are to conceive, how *Tapestry* or *flowred Stuffs* are woven. And the ends of all these Inquiries they intend to be the *Pleasure* of Contemplative minds, but above all, the *ease and dispatch* of the labours of mens hands. They do indeed neglect no opportunity to bring all the *rare* things of Remote Countries within the compass of their knowledge and practice. But they still acknowledge their *most useful* Informations to arise from *common* things, and from *diversifying* their most *ordinary* operations upon them. They do not wholly reject Experiments of meer *light* and *theory*; but they principally aim at such, whose Applications will *improve and facilitate* the present way of *Manual Arts*. And though some men, who are perhaps taken up about less honourable Employments, are pleas'd to censure their proceedings, yet they can shew more *fruits* of their first three years, wherein they have assembled, then any other *Society* in *Europe* can for a much larger space of time. 'Tis true, such undertakings as theirs do commonly meet with small incouragement, because men are generally rather taken with the *plausible* and *discursive*, then the *real* and the solid part of Philosophy; yet by the good fortune of their institution, in an Age of all others the most *inquisitive*, they have been assisted by the *contribution* and *presence* of very many of the chief *Nobility* and *Gentry*, and others, who are some of the *most considerable* in their several Professions. But that that yet farther convinces me of the *Real esteem* that the more *serious* part of men have of this *Society*, is, that several *Merchants*, men who act in earnest (whose Object is *meum* and *tuum*, that great *Rudder* of humane affairs) have adventur'd considerable sums of *Money*, to put in practice what some of our Members have contrived, and have continued *stedfast* in their good opinions of such Indeavours, when not one of a hundred of the vulgar have believed their undertakings feasable. And it is also fit to be added, that they have one advantage peculiar to themselves, that very many of their number are *men of Converse and Traffick*; which is a good Omen, that their attempts will bring Philosophy from *words* to *action*, seeing the men of Business have had so great a share in their first foundation.

And of this kind I ought not to conceal one particular *Generosity*, which more nearly concerns my self. It is the *munificence* of *Sir John Cutler*, in endowing a Lecture for the promotion of *Mechanick Arts*, to be governed and directed by This *Society*. This *Bounty* I mention for the *Honourableness* of the thing it self, and for the expectation which I have of the *efficacy* of the *Example*; for it cannot now be objected to them, that their Designs will be esteemed *frivolous* and *vain*, when they have such a *real Testimony* of the *Approbation* of a *Man* that is such an *eminent Ornament* of this renowned City, and one, who, by the *Variety*, and the *happy Success*, of his negotiations, has given evident proofs, that he is not easie to be deceiv'd. This Gentleman has well observ'd, that the *Arts* of life have been too long *imprison'd* in the dark shops of Mechanicks themselves, and there *hindred from growth*, either by ignorance, or self-interest; and he has bravely *freed* them from these *inconveniences*; He hath not only obliged *Tradesmen*, but *Trade* it self: He has done a work that is worthy of *London*, and has taught the chief City of Commerce in the world the right way how Commerce is to be improv'd. We have already seen many other great signs of Liberality and a large mind, from the same hand: For by his *diligence* about the *Corporation for the Poor;* by his honorable *Subscriptions* for the rebuilding of St. *Paul's*; by his chearful *Disbursment* for the replanting of *Ireland*, and by many other such *publick works*, he has shewn by what means he indeavours to *establish* his Memory; and now by this last gift he has done that, which became one of the *wisest Citizens* of our Nation to accomplish, seeing one of the *wisest of our Statesmen, the Lord Verulam*, first propounded it.

But to return to my Subject, from a digression, which, I hope, my Reader will pardon me, seeing the Example is so rare that I can make no more such digressions. If these my first Labours shall be any wayes useful to inquiring men, I must attribute the incouragement and promotion of them to a very *Reverend* and *Learned Person*, of whom this ought in justice to be said, *That there is scarce any one Invention, which this Nation has produc'd in our Age, but it has some way or other been set forward by his assistance.* My reader, I believe, will quickly ghess, that it is *Dr. Wilkins* that I mean. He is indeed a man born for the *good* of *mankind*, and for the *honour* of his *Country*. In the *sweetness* of whose *behaviour*, in the *calmness* of his *mind*, in the *unbounded goodness* of his *heart*, we have an evident Instance, what the true and the *primitive unpassionate Religion* was, before it was *sowred* by particular *Factions*. In a word, his *Zeal* has been so *constant* and *effectual* in advancing all good and profitable *Arts*, that as one of the Antient *Romans* said of *Scipio, That he thanked God that he was a* Roman; *because whereever* Scipio *had been born, there had been the seat of the Empire of the world:* So may I thank God, that *Dr. Wilkins* was an *Englishman*, for whereever he had lived, there had been the chief Seat of *generous Knowledge* and *true Philosophy*. To the truth of this, there are so many worthy men living that will subscribe, that I am confident, what I have here said, will not be look'd upon, by any ingenious Reader, as a *Panegyrick*, but only as a *real testimony*.

By the Advice of this *Excellent man* I first set upon this Enterprise, yet still came to it with much *Reluctancy*, because I was to follow the footsteps of so eminent a Person as *Dr. Wren*, who was the first that attempted any thing of this nature; whose original draughts do now make one of the Ornaments of that great Collection of Rarities in the *Kings Closet*. This *Honor*, which his first beginnings of this kind have receiv'd, to be admitted into the most famous place of the world, did not so much *incourage*, as the *hazard* of coming after *Dr. Wren* did *affright* me; for of him I must affirm, that, since the time of *Archimedes*, there scarce ever met in one man, in so great a perfection, such a *Mechanical Hand*, and so *Philosophical* a *Mind*.

But at last, being assured both by *Dr. Wilkins*, and *Dr. Wren* himself, that he had given over his intentions of prosecuting it, and not finding that there was any else design'd the pursuing of it, I set upon this undertaking, and was not a little incourag'd to proceed in it, by the Honour the *Royal Society* was pleas'd to favour me with, in approving of those draughts (which from time to time as I had an opportunity of describing) I presented to them. And particularly by the Incitements of divers of those Noble and excellent Persons of it, which were my more especial Friends, who were not less urgent with me for the publishing, then for the prosecution of them.

After I had almost compleated these Pictures and

Observations (having had divers of them ingraven, and was ready to send them to the Press) I was inform'd, that the Ingenious Physitian *Dr. Henry Power* had made several *Microscopical* Observations, which had I not afterwards, upon our interchangably viewing each others Papers, found that they were for the most part differing from mine, either in the Subject it self, or in the particulars taken notice of; and that his design was only to print Observations without Pictures, I had even then *suppressed* what I had so far proceeded in. But being further *excited* by several of my Friends, in complyance with their opinions, that it would not be unacceptable to several inquisitive Men, and hoping also, that I should thereby discover something New to the World, I have at length cast in my Mite, into the vast Treasury of *A Philosophical History*. And it is my *hope*, as well as *belief*, that these my *Labours* will be no more comparable to the *Productions* of many other *Natural Philosophers*, who are now every where busie about *greater* things; then my *little Objects* are to be compar'd to the greater and more beautiful *Works of Nature*, A Flea, a Mite, a Gnat, to an Horse, an Elephant, or a Lyon.

OBSERV. VIII.

*OF THE FIERY SPARKS STRUCK FROM A FLINT OR STEEL.**

IT is a very common Experiment, by striking with a Flint against a Steel, to make certain fiery and shining Sparks to fly out from between those two compressing Bodies. About eight years since, upon casually reading the Explication of this odd *Phænomenon*, by the most Ingenious *Des Cartes*, I had a great desire to be satisfied, what that Subştance was that gave such a shining and bright Light: And to that end I spread a sheet of white Paper, and on it, observing the place where several of these Sparks seemed to vanish, I found certain very small, black, but glistering Spots of a movable Substance, each of which examining with my *Microscope*, I found to be a small round *Globule;* some of which, as they looked prety small, so did they from their Surface yield a very bright and strong reflection on that side which was next the Light; and each look'd almost like a prety bright Iron-Ball, whose Surface was prety regular, such as is represented by the Figure A.† In this I could perceive the Image of the Window prety well, or of a Stick, which I moved up and down between the Light and it. Others I found, which were, as to the bulk of the Ball, prety regularly round, but the Surface of them, as it was not very smooth, but rough, and more irregular, so was the reflection from it more faint and confused. Such were the Surfaces of B. C. D. and E.

* Pp. 44-47. † [The figures have not been reproduced.]

Some of these I found cleft or cracked, as C, others quite broken in two and hollow, as D. which seemed to be half the hollow shell of a Granado, broken irregularly in pieces. Several others I found of other shapes; but that which is represented by E, I observed to be a very big Spark of Fire, which went out upon one side of the Flint that I struck fire withall, to which it stuck by the root F, at the end of which small Stem was fastened-on a *Hemisphere*, or half a hollow Ball, with the mouth of it open from the stemwards, so that it looked much like a Funnel, or an old fashioned Bowl without a foot. This night, making many tryals and observations of this Experiment, I met, among a multitude of the Globular ones which I had observed, a couple of Instances, which are very remarkable to the confirmation of my *Hypothesis.*

And the First was of a pretty big Ball fastened on to the end of a small sliver of Iron, which *Compositum* seemed to be nothing else but a long thin chip of Iron, one of whose ends was melted into a small round Globul; the other end remaining unmelted and irregular, and perfectly Iron.

The Second Instance was not less remarkable then the First; for I found, when a Spark went out, nothing but a very small thin long sliver of Iron or Steel, unmelted at either end. So that it seems, that some of these Sparks are the slivers or chips of the Iron *vitrified*, Others are only the slivers melted into Balls without vitrification. And the third kind are only small slivers of the Iron, made red-hot with the violence of the stroke given on the Steel by the Flint.

He that shall diligently examine the *Phænomena* of this Experiment, will, I doubt not, find cause to believe, that the reason I have heretofore given of it, is the true and genuine cause of it, namely, That *the Spark appear-*

ing so bright in the falling, is nothing else but a small piece of the Steel or Flint, but most commonly of the Steel, which by the violence of the stroke is at the same time sever'd and heatt red-hot, and that sometimes to such a degree, as to make it melt together into a small Globule of Steel; and sometimes also is that heat so very intense, as further to melt it and vitrifie it; but many times the heat is so gentle, as to be able to make the sliver only red hot, which notwithstanding falling upon the tinder (that is only a very curious small Coal made of the small threads of Linnen burnt to coals and char'd) *it easily sets it on fire.* Nor will any part of this *Hypothesis* seem strange to him that considers, First, that either hammering, or filing, or otherwise violently rubbing of Steel, will presently make it so hot as to be able to burn ones fingers. Next, that the whole force of the stroke is exerted upon that small part where the Flint and Steel first touch: For the Bodies being each of them so very hard, the puls cannot be far communicated, that is, the parts of each can yield but very little, and therefore the violence of the concussion will be *exerted* on that piece of Steel which is cut off by the Flint. Thirdly, that the filings or small parts of Steel are very apt, as it were, to take fire, and are presently red hot, that is, there seems to be a very *combustible sulphureous* Body in Iron or Steel, which the Air very readily preys upon, as soon as the body is a little violently heated.

And this is obvious in the filings of Steel or Iron cast through the flame of a Candle; for even by that sudden *transitus* of the small chips of Iron, they are heat red hot, and that *combustible sulphureous* Body is presently prey'd upon and devoured by the *aereal* incompassing *Menstruum*, whose office in this Particular I have shewn in the Explication of Charcole.

And in prosecution of this Experiment, having taken

the filings of Iron and Steel, and with the point of a Knife cast them through the flame of a Candle, I observed where some conspicuous shining Particles fell, and looking on them with my *Microscope*, I found them to be nothing else but such round Globules, as I formerly found the Sparks struck from the Steel by a stroke to be, only a little bigger; and shaking together all the filings that had fallen upon the sheet of Paper underneath, and observing them with the *Microscope*, I found a great number of small Globules, such as the former, though there were also many of the parts that had remained untoucht, and rough filings or chips of Iron. So that, it seems, Iron does contain a very *combustible sulphureous* Body, which is, in all likelihood, one of the causes of this *Phænomenon*, and which may be perhaps very much concerned in the business of its hardening and tempering: of which somewhat is said in the Description of *Muscovy-glass*.

So that, these things considered, we need not trouble our selves to find out what kind of Pores they are, both in the Flint and Steel, that contain the *Atoms of fire*, nor how those *Atoms* come to be hindred from running all out, when a dore or passage in their Pores is made by the concussion: nor need we trouble our selves to examine by what *Prometheus* the Element of Fire comes to be fetcht down from above the Regions of the Air, in what Cells or Boxes it is kept, and what *Epimetheus* lets it go: Nor to consider what it is that causes so great a conflux of the atomical Particles of Fire, which are said to fly to a flaming Body, like Vultures or Eagles to a putrifying Carcass, and there to make a very great pudder. Since we have nothing more difficult in this *Hypothesis* to conceive, first, as to the kindling of Tinder, then how a large Iron-bullet, let fall red or glowing hot upon a heap of Small-coal, should set fire to those that are next to it first: Nor secondly, is this last more difficult to be explicated, then that a Body, as Silver for Instance, put into a weak *Menstruum*, as unrectified *Aqua fortis* should, when it is put in a great heat, be there dissolved by it, and not before; which *Hypothesis* is more largely explicated in the Description of Charcoal. To conclude, we see by this Instance, how much Experiments may conduce to the regulating of *Philosophical notions*. For if the most Acute *Des Cartes* had applied himself experimentally to have examined what substance it was that caused that shining of the falling Sparks struck from a Flint and a Steel, he would certainly have a little altered his *Hypothesis*, and we should have found, that his Ingenious Principles would have admitted a very plausible Explication of this *Phænomenon;* whereas by not examining so far as he might, he has set down an Explication which Experiment do's contradict.

But before I leave this Description, I must not forget to take notice of the Globular form into which each of these is most curiously formed. And this *Phænomenon*, as I have elsewhere more largely shewn, proceeds from a propriety which belongs to all kinds of fluid Bodies more or less, and is caused by the Incongruity of the Ambient and included Fluid, which so acts and modulates each other, that they acquire, as neer as is possible, a *spherical* or *globular* form, which propriety and several of the *Phænomena* that proceed from it, I have more fully explicated in the sixth Observation.

One Experiment, which does very much illustrate my present Explication, and is in it self exceeding pretty, I must not pass by: And that is a way of making small *Globules* or *Balls* of Lead, or Tin, as small almost as these of Iron or Steel, and that exceeding easily and quickly, by turning the filings or chips of those Metals also into perfectly round *Globules*. The way, in short, as I received it from the *Learned Physitian Doctor* I. G. is this;

Reduce the Metal you would thus shape, into exceeding fine filings, the finer the filings are, the finer will the Balls be: *Stratifie* these filings with the fine and well dryed powder of quick Lime in a *Crucible* proportioned to the quantity you intend to make: When you have thus filled your *Crucible*, by continual *stratifications* of the filings and powder, so that, as neer as may be, no one of the filings may touch another, place the *Crucible* in a *gradual fire*, and by degrees let it be brought to a heat big enough to make all the filings, that are mixt with the quick Lime, to melt, and no more; for if the fire be too hot, many of these filings will joyn and run together; whereas if the heat be proportioned, upon washing the Lime-dust in fair Water, all those small filings of the Metal will subside to the bottom in a most curious powder, consisting all of exactly round *Globules*, which, if it be very fine, is very excellent to make Hour-glasses of.

Now though quick Lime be the powder that this direction makes choice of, yet I doubt not, but that there may be much more convenient ones found out, one of which I have made tryal of, and found very effectual; and were it not for discovering, by the mentioning of it, another Secret, which I am not free to impart, I should have here inserted it.

EXTRACTS FROM OBSERV. IX.

*OF THE COLOURS OBSERVABLE IN MUSCOVY GLASS, AND OTHER THIN BODIES.**

THOSE several Colours which are observed to follow each other upon the polisht surface of hardned Steel, when it is by a sufficient degree of heat gradually tempered or softened, are produced from nothing else but a certain thin *Lamina* of a *vitrum* or *vitrified* part of the Metal, which by that degree of heat, and the concurring action of the ambient Air, is driven out and fixed on the surface of the Steel.

And this hints to me a very probable (at least, if not the true) cause of the hardning and tempering of Steel, which has not, I think, been yet given, nor, that I know of, been so much as thought of by any. And that is this, that the hardness of it arises from a greater proportion of a vitrified Substance interspersed through the pores of the Steel. And that the tempering or softning of it arises from the proportionate or smaller parcels of it left within those pores. This will seem the more probable, if we consider these Particulars.

First, That the pure parts of Metals are of themselves very *flexible* and *tuff;* that is, will indure bending and hammering, and yet retain their continuity.

Next, That the Parts of all vitrified Substances, as all kinds of Glass, the *Scoria* of Metals, *&c.* are very hard, and also very brittle, being neither *flexible* nor *malleable*, but may by hammering or beating be broken into small parts or powders.

---

* Pp. 51-53 and p. 55.

Thirdly, That all Metals (excepting Gold and Silver, which do not so much with the bare fire, unless assisted by other saline Bodies) do more or less *vitrifie* by the strength of fire, that is, are corroded by a saline Substance, which I elsewhere shew to be the true cause of fire; and are thereby, as by several other *Menstruums*, converted into *Scoria;* And this is called, *calcining* of them, by Chimists. Thus Iron and Copper by heating and quenching do turn all of them by degrees into *Scoria*, which are evidently *vitrified* Substances, and unite with Glass, and are easily *fusible;* and when cold, very hard, and very brittle.

Fourthly, That most kind of *Vitrifications* or *Calcinations* are made by Salts, uniting and incorporating with the metalline Particles. Nor do I know any one *calcination* wherein a *Saline* body may not, with very great probability, be said to be an agent or coadjutor.

Fifthly, That Iron is converted into Steel by means of the incorporation of certain salts, with which it is kept a certain time in the fire.

Sixthly, That any Iron may, in a very little time, be *case hardned*, as the Trades-men call it, by casing the iron to be hardned with clay, and putting between the clay and iron a good quantity of a mixture of *Urine*, *Soot*, *Sea-salt*, and *Horses hoofs* (all which contein great quantities of Saline bodies) and then putting the case into a good strong fire, and keeping it in a considerable degree of heat for a good while, and afterwards heating, and quenching or cooling it suddenly in cold water.

Seventhly, That all kind of vitrify'd substances, by being suddenly cool'd, become very hard and brittle. And thence arises the pretty *Phænomena* of the Glass Drops, which I have already further explained in its own place.

Eighthly, That those metals which are not so apt to

vitrifie, do not acquire any hardness by quenching in water, as Silver, Gold, &c.

These considerations premis'd, will, I suppose, make way for the more easie reception of this following Explication of the *Phænomena* of hardned and temper'd Steel. That Steel is a substance made out of iron, by means of a certain proportionate *Vitrification* of several parts, which are so curiously and proportionately mixt with the more tough and unalter'd parts of the Iron, that when by the great heat of the fire this vitrify'd substance is melted, and consequently rarify'd, and thereby the pores of the Iron are more open; if then by means of dipping it in cold water it be suddenly cold, and the parts hardned, that is, stay'd in that same degree of *Expansion* they were in when hot, the parts become very hard and brittle, and that upon the same account almost as small parcels of glass quenched in water grow brittle, which we have already explicated. If after this the piece of Steel be held in some convenient heat, till by degrees certain colours appear upon the surface of the brightned metal, the very hard and brittle tone of the metal, by degrees relaxes and becomes much more tough and soft; namely, the action of the heat does by degrees loosen the parts of the Steel that were before streached or set *atilt* as it were, and stayed open by each other, whereby they become relaxed and set at liberty, whence some of the more brittle interjacent parts are thrust out and melted into a thin skin on the surface of the Steel, which from no colour increases to a deep Purple, and so onward by these *gradations* or consecutions, *White*, *Yellow*, *Orange*, *Minium*, *Scarlet*, *Purple*, *Blew*, *Watchet*, &c. and the parts within are more conveniently, and proportionately mixt; and so they gradually subside into a texture which is much better proportion'd and closer joyn'd, whence that rigidnesse of parts ceases, and the parts begin to acquire their former *ductilness*.

Now, that 'tis nothing but the vitrify'd metal that sticks upon the surface of the colour'd body, is evident from this, that if by any means it be scraped and rubb'd off, the metal underneath it is white and clear; and if it be kept longer in the fire, so as to increase to a considerable thickness, it may, by blows, be beaten off in flakes. This is further confirm'd by this observable, that that Iron or Steel will keep longer from rusting which is covered with this vitrify'd case: Thus also Lead will, by degrees, be all turn'd into a litharge; for that colour which covers the top being scum'd or shov'd aside, appears to be nothing else but a litharge or vitrify'd Lead.

This is observable also in some sort, on Brass, Copper, Silver, Gold, Tin, but is most conspicuous in Lead: all those Colours that cover the surface of the Metal being nothing else, but a very thin vitrifi'd part of the heated Metal.

* * *

It would be too long, I say, here to insert the discursive progress by which I inquir'd after the proprieties of the motion of Light, and therefore I shall only add the result.

And, First, I found it ought to be exceeding *quick*, such as those motions of *fermentation* and *putrefaction*, whereby, certainly, the parts are exceeding nimbly and violently mov'd; and that, because we find those motions are able more minutely to shatter and divide the body, then the most violent heats or *menstruums* we yet know. And that fire is nothing else but such a *dissolution* of the Burning body, made by the most *universal menstruum* of all *sulphureous bodies*, namely, the Air, we shall in an other place of this Tractate endeavour to make probable. And that, in all extreamly hot shining bodies, there is a very quick motion that causes Light, as well as a more robust that causes Heat, may be argued from the celerity wherewith the bodyes are dissolv'd.

OBSERV. XVI.

OF CHARCOAL, OR *BURNT* VEGETABLES.*

CHARCOAL, or a Vegetable burnt black, affords an object no less pleasant than instructive; for if you take a small round Charcoal, and break it short with your fingers, you may perceive it to break with a very smooth and sleek surface, almost like the surface of black sealing Wax; this surface, if it be look'd on with an ordinary *Microscope*, does manifest abundance of those pores which are also visible to the eye in many kinds of *Wood*, rang'd round the pith, both in a kind of circular order, and a radiant one. Of these there are a multitude in the substance of the Coal, every where almost perforating and drilling it from end to end; by means of which, be the Coal never so long, you may easily blow through it; and this you may presently find, by wetting one end of it with Spittle, and blowing at the other.

But this is not all, for besides those many great and conspicuous irregular spots or pores, if a better *Microscope* be made use of, there will appear an infinite company of exceedingly small, and very regular pores, so thick and so orderly set, and so close to one another, that they leave very little room or space between them to be fill'd with a solid body, for the apparent *interstitia*, or separating sides of these pores seem so thin in some places, that the texture of a Honey-comb cannot be more porous. Though this be not every where so, the intercurrent

---

* Pp. 100-106.

partitions in some places being very much thicker in proportion to the holes.

Most of these small pores seem'd to be pretty round, and were rang'd in rows that radiated from the pith to the bark; they all of them seem'd to be continued open pores, running the whole length of the Stick; and that they were all perforated, I try'd by breaking off a very thin sliver of the Coal cross-ways, and then with my *Microscope*, diligently surveying them against the light, for by that means I was able to see quite through them.

These pores were so exceeding small and thick, that in a line of them, $\frac{1}{18}$ part of an Inch long, I found by numbring them no less then 150. small pores; and therefore in a line of them an Inch long, must be no less then 2700. pores, and in a circular *area* of an Inch diameter, must be about 5725350. of the like pores; so that a Stick of an Inch Diameter, may containe no less then seven hundred and twenty five thousand, besides 5 Millions of pores, which would, I doubt not, seem even incredible, were not every one left to believe his own eyes. Nay, having since examin'd *Cocus*, *black and green Ebony*, *Lignum Vitæ*, &c. I found, that all these Woods have their pores, abundantly smaller then those of soft light Wood; in so much, that those of *Guajacum* seem'd not above an eighth part of the bigness of the pores of Beech, but then the *Interstitia* were thicker; so prodigiously curious are the contrivances, pipes, or sluces by which the *Succus nutritius*, or Juyce of a Vegetable is convey'd from place to place.

This *Observation* seems to afford us the true reason of several *Phænomena* of Coals; as

First, why they look black; and for this we need go no further then the *Scheme*, for certainly, a body that has so many pores in it as this is discover'd to have, from each of which no light is reflected, must necessarily look black, especially, when the pores are somewhat bigger in proportion to the intervals then they are cut in the *Scheme*, black being nothing else but a privation of Light, or a want of reflection ; and wheresoever this reflecting quality is deficient, there does that part look black, whether it be from a porousness of the body, as in this Instance, or in a deadning and dulling quality, such as I have observ'd in the *Scoria* of Lead, Tin, Silver, Copper, *&c.*

Next, we may also as plainly see the reason of its shining quality, and that is from the even breaking off of the stick, the solid *interstitia* having a regular termination or surface, and having a pretty strong reflecting quality, the many small reflections become united to the naked eye, and make a very pretty shining surface.

Thirdly, the reason of its hardness and brittleness seems evident, for since all the watery or liquid substance that moistn'd and toughn'd those *Interstitia* of the more solid parts, are evaporated and remov'd, that which is left behind becomes of the nature almost of a stone, which will not at all, or very little, bend without a *divulsion* or *solution* of its *continuity.*

It is not my design at present, to examine the use and *Mechanisme* of these parts of Wood, that being more proper to another Enquiry ; but rather to hint, that from this Experiment we may learn,

First, what is the cause of the blackness of many burnt bodies, which we may find to be nothing else but this ; that the heat of the fire agitating and rarifying the waterish, transparent, and volatile water that is contain'd in them, by the continuation of that action, does so totally expel and drive away all that which before fill'd the pores, and was dispers'd also through the solid mass of it, and thereby caus'd an universal kind of transparency, that it not onely leaves all the pores empty, but all the *Interstitia* also so dry and *opacous*, and perhaps also yet further perforated, that that light onely is reflected back which falls upon the very outward edges of the pores, all they that enter into the pores of the body, never returning, but being lost in it.

Now, that the Charring or coaling of a body is nothing else, may be easily believ'd by one that shall consider the means of its production, which may be done after this, or any such manner. The body to be charr'd or coal'd, may be put into a *Crucible*, Pot, or any other Vessel that will endure to be made red-hot in the Fire without breaking, and then cover'd over with Sand, so as no part of it be suffer'd to be open to the Air, then set into a good Fire, and there kept till the Sand has continu'd red hot for a quarter, half, an hour or two, or more, according to the nature and bigness of the body to be coal'd or charr'd, then taking it out of the Fire, and letting it stand till it be quite cold, the body may be taken out of the Sand well charr'd and cleans'd of its waterish parts; but in the taking of it out, care must be had that the Sand be very neer cold, for else, when it comes into the free air, it will take fire, and readily burn away.

This may be done also in any close Vessel of Glass, as a *Retort*, or the like, and the several fluid substances that come over may be receiv'd in a fit *Recipient*, which will yet further countenance this *Hypothesis*: And their manner of charring Wood in great quantity comes much to the same thing, namely, an application of a great heat to the body, and preserving it from the free access of the devouring air; this may be easily learn'd from the History of Charring of Coal, most excellently describ'd and publish'd by that most accomplish'd Gentleman, Mr *John Evelin*, in the 100, 101, 103, pages of his *Sylva*, to which I shall therefore refer the curious Reader that desires a full information of it.

Next, we may learn what part of the Wood it is that is the *combustible* matter; for since we shall find that none, or very little of those fluid substances that are driven over into the Receiver are *combustible*, and that most of that which is left behind is so, it follows, that the solid *interstitia* of the Wood are the *combustible* matter. Further, the reason why uncharr'd Wood burns with a greater flame then that which is charr'd, is as evident, because those waterish or volatil parts issuing out of the fired Wood, every way, not onely shatter and open the body, the better for the fire to enter, but issuing out in vapours or wind, they become like so many little *æolipiles*, or Bellows, whereby they blow and agitate the fir'd part, and conduce to the more speedy and violent consumption or dissolution of the body.

Thirdly, from the Experiment of charring of Coals (whereby we see that notwithstanding the great heat, and the duration of it, the solid parts of the Wood remain, whilest they are preserv'd from the free access of the air undissipated) we may learn, that which has not, that I know of, been publish'd or hinted, nay, not so much as thought of, by any; and that in short is this.

First, *that the Air* in which we live, move, and breath, and which encompasses very many, and cherishes most bodies it encompasses, that this Air is the *menstruum*, or universal dissolvent of all *Sulphureous* bodies.

Secondly, *that this action* it performs not, till the body be first sufficiently heated, as we find requisite also to the dissolution of many other bodies by several other *menstruums*.

Thirdly, *that this action* of dissolution, produces or generates a very great heat, and that which we call Fire; and this is common also to many dissolutions of other bodies, made by *menstruums*, of which I could give multitudes of Instances.

Fourthly, *that this action* is perform'd with so great a violence, and does so minutely act, and rapidly agitate the smallest parts of the *combustible* matter, that it produces in the *diaphanous medium* of the Air, the action or pulse of light, which what it is, I have else-where already shewn.

Fifthly, *that the dissolution* of sulphureous bodies is made by a substance inherent, and mixt with the Air, that is like, if not the very same, with that which is fixt in *Salt-peter*, which by multitudes of Experiments that may be made with *Saltpeter*, will, I think, most evidently be demonstrated.

Sixthly, *that in this dissolution* of bodies by the Air, a certain part is united and mixt, or dissolv'd and turn'd into the Air, and made to fly up and down with it in the same manner as a *metalline* or other body dissolv'd into any *menstruums*, does follow the motions and progresses of that *menstruum* till it be precipitated.

Seventhly, That as there is one part that is dissoluble by the Air, so are there other parts with which the parts of the Air mixing and uniting, do make a *Coagulum*, or *precipitation*, as one may call it, which causes it to be separated from the Air, but this *precipitate* is so light, and in so small and rarify'd or porous clusters, that it is very volatil, and is easily carry'd up by the motion of the Air, though afterwards, when the heat and agitation that kept it rarify'd ceases, it easily condenses, and commixt with other indissoluble parts, it sticks and adheres to the next bodies it meets withall; and this is a certain *Salt* that may be extracted out of *Soot*.

Eighthly, that many indissoluble parts being very apt and prompt to be rarify'd, and so, whilest they continue in that heat and agitation, are lighter than the Ambient Air, are thereby thrust and carry'd upwards with great violence, and by that means carry along with them, not onely that *Saline concrete* I mention'd before, but many terrestrial, or indissoluble and irrarefiable parts, nay, many parts also which are dissoluble, but are not suffer'd to stay long enough in a sufficient heat to make them prompt and apt for that action. And therefore we find in *Soot*, not onely a part, that being continued longer in a competent heat, will be dissolv'd by the Air, or take fire and burn; but a part also which is fixt, terrestrial, and irrarefiable.

Ninthly, that as there are these several parts that will rarifie and fly, or be driven up by the heat, so are there many others, that as they are indissoluble by the *aerial menstruum*, so are they of such sluggish and gross parts, that they are not easily rarify'd by heat, and therefore cannot be rais'd by it; the volatility or fixtness of a body seeming to consist only in this, that the one is of a texture, or has component parts that will be easily rarify'd into the form of Air, and the other, that it has such as will not, without much ado, be brought to such a constitution; and this is that part which remains behind in a white body call'd Ashes, which contains a substance, or *Salt*, which Chymists call *Alkali*: what the particular natures of each of these bodies are, I shall not here examine, intending it in another place, but shall rather add that this *Hypothesis* does so exactly agree with all *Phænomena* of Fire, and so genuinely explicate each particular circumstance that I have hitherto observ'd, that it is more then probable, that this cause which I have assign'd is the true adequate, real, and onely cause of those *Phænomena*; And therefore I shall proceed a little further, to shew the nature and use of the Air.

Tenthly, therefore the dissolving parts of the Air are but few, that is, it seems of the nature of those *Saline menstruums*, or spirits, that have very much flegme mixt with the spirits, and therefore a small parcel of it is quickly glutted, and will dissolve no more; and therefore unless some fresh part of this *menstruum* be apply'd to the body to be dissolv'd, the action ceases, and the body leaves to be dissolv'd and to shine, which is the Indication of it, though plac'd or kept in the greatest heat; whereas *Salt-peter* is a *menstruum*, when melted and red-hot, that abounds more with those Dissolvent particles, and therefore as a small quantity of it will dissolve a great sulphureous body, so will the dissolution be very quick and violent.

Therefore in the *Eleventh* place, it is observable, that, as in other solutions, if a copious and quick supply of fresh *menstruum*, though but weak, be poured on, or applied to the dissoluble body, it quickly consumes it: So this *menstruum* of the Air, if by Bellows, or any other such contrivance, it be copiously apply'd to the shining body, is found to dissolve it as soon, and as violently as the more strong *menstruum* of melted *Nitre*.

Therefore twelfthly, it seems reasonable to think that there is no such thing as an Element of Fire that should attract or draw up the flame, or towards which the flame should endeavour to ascend out of a desire or appetite of uniting with that as its *Homogeneal* primitive and generating Element; but that that shining transient body which we call *Flame*, is nothing else but a mixture of Air, and volatil sulphureous parts of dissoluble or combustible bodies, which are acting upon each other whil'st they ascend, that is, flame seems to be a mixture of Air, and the combustible volatil parts of any body, which parts the encompassing Air does dissolve or work upon, which action, as it does intend the heat of the *aerial* parts of the dissolvent, so does it thereby further rarifie those parts that are acting, or that are very neer them, whereby they growing much lighter then the heavie parts of that *Menstruum* that are more remote, are thereby protruded and driven upward; and this may be easily observ'd also in dissolutions made by any other *menstruum*, especially such as either create heat or bubbles. Now, this action of the *Menstruum*, or *Air*, on the dissoluble parts, is made with such violence, or is such, that it imparts such a motion or pulse to the *diaphanous* parts of the Air, as I have elsewhere shewn is requisite to produce light.

This *Hypothesis* I have endeavoured to raise from an Infinite of Observations and Experiments, the process of which would be much too long to be here inserted, and will perhaps another time afford matter copious enough for a much larger Discourse, the Air being a Subject which (though all the world has hitherto liv'd and breath'd in, and been unconversant about) has yet been so little truly examin'd or explain'd, that a diligent enquirer will be able to find but very little information from what has been (till of late) written of it: But being once well understood, it will, I doubt not, inable a man to render an intelligible, nay probable, if not the true reason of all the *Phænomena* of Fire, which, as it has been found by Writers and Philosophers of all Ages a matter of no small difficulty, as may be sufficiently understood by their strange *Hypotheses*, and unintelligible Solutions of some few *Phænomena* of it; so will it prove a matter of no small concern and use in humane affairs, as I shall elsewhere endeavour to manifest when I come to show the use of the Air in respiration, and for the preservation of the life, nay, for the conservation and restauration of the health and natural constitution of mankind as well as all other aereal *animals*, as also the uses of this principle or propriety of the Air in chymical, mechanical, and other operations. In this place I have onely time to hint an *Hypothesis*, which, if God permit me life and opportunity, I may elsewhere prosecute, improve and publish. In the mean time, before I finish this Discourse, I must not forget to acquaint the Reader, that having had the liberty granted me of making some trials on a piece of *Lignum fossile* shewn to the Royal Society, by the eminently Ingenious and Learned Physician, Doctor *Ent*, who receiv'd it for a Present from the famous *Ingenioso Cavalliero de Pozzi*, it being one of the fairest and best pieces of *Lignum fossile* he had seen; Having (I say) taken a small piece of this Wood, and examin'd it, I found it to burn in the open Air almost like other Wood, and insteed of a resinous smoak or fume, it yielded a very bituminous one, smelling much of that kind of sent: But that which I chiefly took notice of, was, that cutting off a small piece of it, about the bigness of my Thumb, and charring it in a *Crucible* with Sand, after the manner I above prescrib'd, I found it infinitely to abound with the smaller sort of pores, so extreamly thick, and so regularly perforating the substance of it long-ways, that breaking it off a-cross, I found it to look very like an Honey-comb; but as for any of the second, or bigger kind of pores, I could not find that it had any; so that it seems, whatever were the cause of its production, it was not without those small kind of pores which we have onely hitherto found in Vegetable bodies: and comparing them with the pores which I have found in the Charcoals that I by this means made of several other kinds of Wood, I find it resemble none so much as those of Firr, to which it is not much unlike in grain also, and several other proprieties.

And therefore, what ever is by some, who have written of it, and particularly by *Francisco Stelluto*, who wrote a Treatise in *Italian* of that Subject, which was Printed at *Rome*, 1637. affirm'd that it is a certain kind of Clay or Earth, which in tract of time is turn'd into Wood, I rather suspect the quite contrary, that it was at first certain great Trees of Fir or Pine, which by some Earth-quake, or other casualty, came to be buried under the Earth, and was there, after a long time's residence (according to the several natures of the encompassing adjacent parts) either rotted and turn'd into a kind of Clay, or *petrify'd* and turn'd into a kind of Stone, or else had its pores fill'd with certain Mineral juices, which being stayd in them, and in tract of time coagulated, appear'd, upon cleaving out, like small Metaline Wires, or else from some flames or scorching forms that are the occasion oftentimes, and usually accompany Earthquakes, might be blasted and turn'd into Coal, or else from certain *subterraneous* fires which are affirm'd by that Authour to abound much about those parts (namely, in a Province of *Italy*, call'd *Umbria*, now the *Dutchie* of *Spoletto*, in the Territory of *Todi*, anciently call'd *Tudor*; and between the two villages of *Collesecco* and *Rosaro* not far distant from the high-way leading to *Rome*, where it is found in greater quantity than elsewhere) are by reason of their being encompassed with Earth, and so kept close from the dissolving Air, charr'd and converted into Coal. It would be too long a work to describe the several kinds of pores which I met withall, and by this means discovered in several other Vegetable bodies; nor is it my present design to expatiate upon Instances of the same kind, but rather to give a Specimen of as many kinds as I have had opportunity as yet of observing, reserving the prosecution and enlarging on particulars till a more fit opportunity.

Extracts from Observ. VI.

*OF SMALL GLASS CANES.**

AND we know, that a sufficient degree of heat causes *fluidity*, in some bodies much sooner, and in others later; that is, the parts of the body of some are so *loose* from one another, and so *unapt to cohere*, and so *minute* and *little*, that a very *small* degree of agitation keeps them always in the *state of fluidity*. Of this kind, I suppose, the *Æther*, that is the *medium* or *fluid* body, in which all other bodies do as it were swim and move; and particularly, the *Air*, which seems nothing else but a kind of *tincture* or *solution* of terrestrial and aqueous particles *dissolv'd* into it, and agitated by it, just as the *tincture* of *Cocheneel* is nothing but some finer *dissoluble* parts of that Concrete lick'd up or *dissolv'd* by the *fluid* water.

* * *

For I neither conclude from one single Experiment, nor are the Experiments I make use of, all made upon one Subject: Nor wrest I any Experiment to make it *quadrare* with any preconceiv'd Notion. But on the contrary, I endeavour to be conversant in divers kinds of Experiments, and all and every one of those Trials, I make the Standards or Touchstones, by which I try all my former Notions, whether they hold out in weight, and measure, and touch, *&c.* For as that Body is no other then a Counterfeit Gold, which wants any one of the Proprieties of Gold, (such as are the Malleableness, Weight, Colour, Fixtness in the Fire, Indissolubleness in *Aqua fortis*, and the like) though it has all the other; so will all those Notions be found to be false and deceitful, that

* Pp. 13 and 28.

will not undergo all the Trials and Tests made of them by Experiments. And therefore such as will not come up to the desired *Apex* of Perfection, I rather wholly reject and take new, then by piecing and patching, endeavour to retain the old, as knowing such things at best to be but lame and imperfect. And this course I learned from Nature; whom we find neglectful of the old Body, and suffering its Decaies and Infirmities to remain without repair, and altogether sollicitous and careful of perpetuating the *Species* by new *Individuals.* And it is certainly the most likely way to erect a glorious Structure and Temple to *Nature*, such as she will be found (by any *zealous Votary*) to reside in; to begin to build a new upon a sure Foundation of Experiments.

Extract from Observ. XXII.

Of Common Sponges, *and Several Other* Spongie *Fibrous Bodies.**

I am very apt to think, that were there a contrivance whereby the expir'd air might be forc'd to pass through a wet or oyly Sponge before it were again inspir'd, it might much cleanse, and strain away from the Air divers fuliginous and other noisome steams, and the dipping of it in certain liquors might, perhaps, so renew that property in the Air which it loses in the Lungs, by being breath'd, that one square foot of Air might last a man for respiration much longer, perhaps, then ten will now serve him of common Air.

* P. 140.

EXTRACT FROM OBSERV. LII.

OF THE SMALL SILVER-COLOUR'D BOOK-WORM.*

THIS Animal probably feeds upon the Paper and covers of Books, and perforates in them several small round holes, finding, perhaps, a convenient nourishment in those husks of Hemp and Flax, which have pass'd through so many scourings, washings, dressings and dryings, as the parts of old Paper must necessarily have suffer'd; the digestive faculty, it seems, of these little creatures being able yet further to work upon those stubborn parts, and reduce them into another form.

And indeed, when I consider what a heap of Saw-dust or chips this little creature (which is one of the teeth of Time) conveys into its intrals, I cannot chuse but remember and admire the excellent contrivance of Nature, in placing in Animals such a fire, as is continually nourished and supply'd by the materials convey'd into the stomach, and *fomented* by the bellows of the lungs; and in so contriving the most admirable fabrick of Animals, as to make the very spending and wasting of that fire, to be instrumental to the procuring and collecting more materials to augment and cherish it self, which indeed seems to be the principal end of all the contrivances observable in bruit Animals.

---

* Pp. 209-210.

LaVergne, TN USA
08 December 2010
207949LV00002B/145/A

9 781104 126308